By Sindy McKay

WE BOTH READ®

Parent's Introduction

Whether your child is a beginning reader, a reluctant reader, or an eager reader, this book offers a fun and easy way to encourage and help your child in reading.

Developed with reading education specialists, **We Both Read** books invite you and your child to take turns reading aloud. You read the left-hand pages of the book, and your child reads the right-hand pages—which have been written at one of six early reading levels. The result is a wonderful new reading experience and faster reading development!

You may find it helpful to read the entire book aloud yourself the first time, then invite your child to participate the second time. As you read, try to make the story come alive by reading with expression. This will help to model good fluency. It will also be helpful to stop at various points to discuss what you are reading. This will help increase your child's understanding of what is being read.

In some books, a few challenging words are introduced in the parent's text, distinguished with **bold** lettering. Pointing out and discussing these words can help to build your child's reading vocabulary. If your child is a beginning reader, it may be helpful to run a finger under the text as each of you reads. Please also notice that a "talking parent" icon precedes the parent's text, and a "talking child" icon precedes the child's text.

If your child struggles with a word, you can encourage "sounding it out," but keep in mind that not all words can be sounded out. Your child might pick up clues about a word from the picture, other words in the sentence, or any rhyming patterns. If your child struggles with a word for more than five seconds, it is usually best to simply say the word.

Most of all, remember to praise your child's efforts and keep the reading fun. At the end of the book, there is a glossary of words, as well as some questions you can discuss. Rereading this book multiple times may also be helpful for your child.

Try to keep the tips above in mind as you read together, but don't worry about doing everything right. Simply sharing the enjoyment of reading together will increase your child's reading skills and help to start your child off on a lifetime of reading enjoyment!

The Ocean

A We Both Read Book
Level 1–2
Guided Reading: Level H

With special thanks to
Rebecca Albright, Ph.D.,
Curator, California Academy of Sciences,
for her advice on the material in this book

To Bonnie and Jeremy —and all who come after them.
— S. M.

Use of photographs provided by iStock, Dreamstime, PhotoDisc, and Corbis Images.

Published by
Treasure Bay, Inc.
PO Box 519
Roseville, CA 95661 USA

Printed in China

Library of Congress Catalog Card Number: 2018904179

ISBN: 978-1-60115-310-4

Visit us online at WeBothRead.com

PR-8-21

Table of Contents

CHAPTER 1

OUR BEAUTIFUL OCEANS

Imagine you are a space alien flying high above **Earth**. You look down at the beautiful planet below, and what do you see? Water! You see lots and lots of water. Most of that water is contained in the **oceans** and seas of **Earth**.

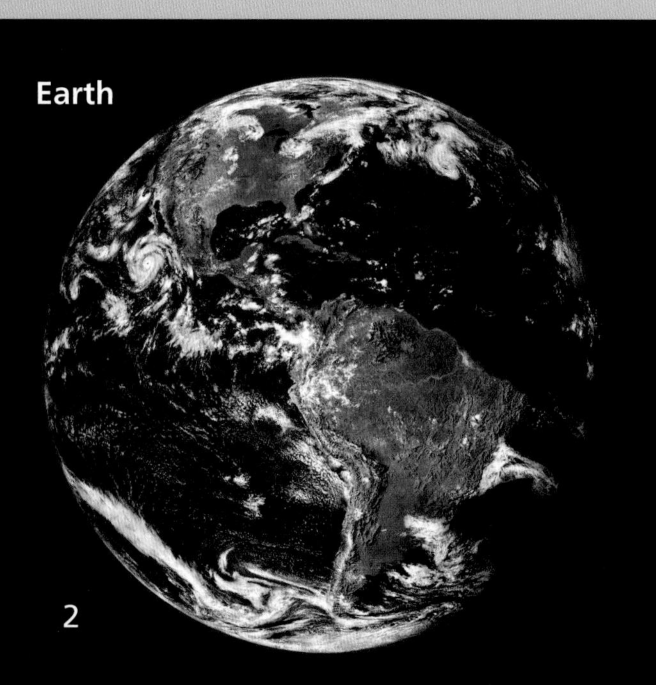

Earth

Earth is covered by much more water than land. You could fit all the land on Earth into the **oceans** more than two times!

There are five major oceans on Earth. They are the **Pacific**, Atlantic, Indian, **Southern,** and Arctic. There are also many smaller seas.

These oceans and most seas are all really one vast worldwide ocean that is broken up by big pieces of land we call *continents.*

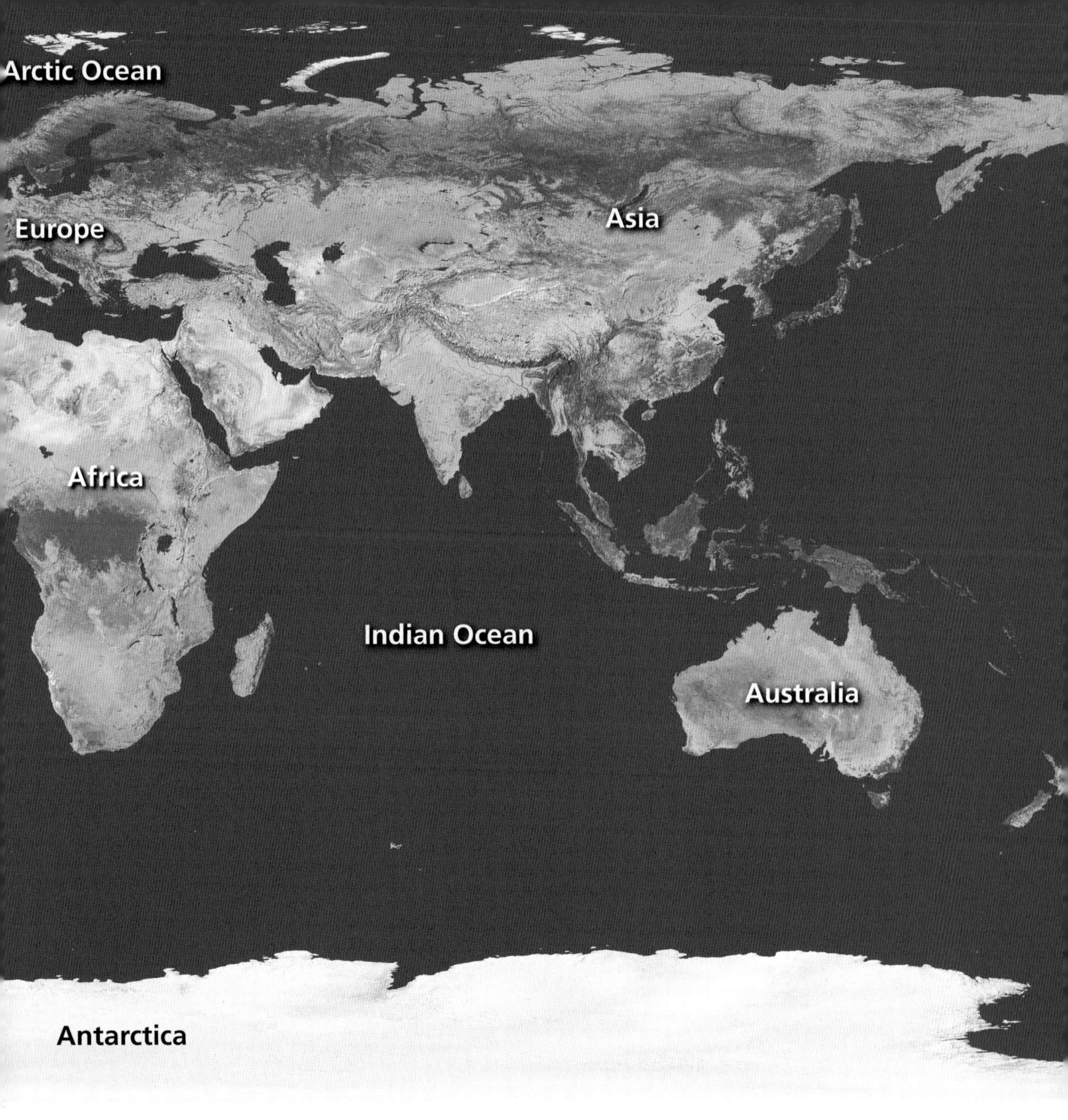

The **Pacific** is the biggest ocean. You can sail on it for many days and never see land. The **Southern** Ocean is the part of the world's ocean that is closest to the South Pole.

When we look out at the sea, we see a vast expanse of water. But when we look under the ocean's surface, we find an amazing world filled with deep trenches, high mountains, dark caves, and colorful coral reefs. We also find an enormous variety of plants and animals from the tiny krill to the mighty whale.

Underwater cave

Longsnout seahorse

Krill swarm

Antarctic krill

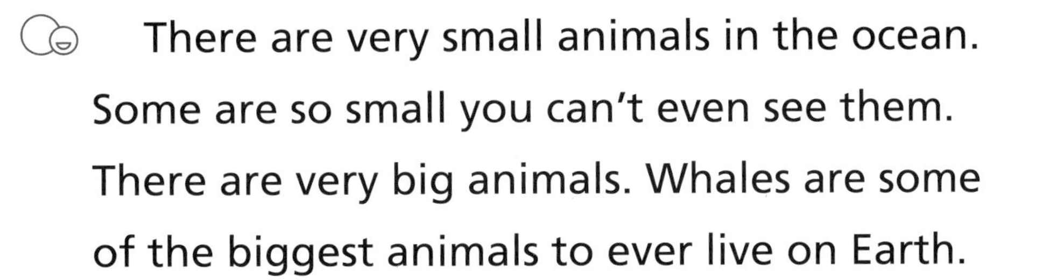

There are very small animals in the ocean. Some are so small you can't even see them. There are very big animals. Whales are some of the biggest animals to ever live on Earth.

Humpback whale

Algae floating in the ocean

Protozoa plankton (seen though a microscope)

Life in the ocean can be divided into three major groups. The first group, called ***plankton***, includes the plants and animals that move and drift with the currents and tides. Some types of algae (AL-jee) are **plankton** that drift in the ocean. Other types of algae cling to the bottom of shallow waters, like tide pools.

Some types of **plankton** can be large, like jellyfish with their long **tentacles**. However, most types of **plankton** are very tiny.

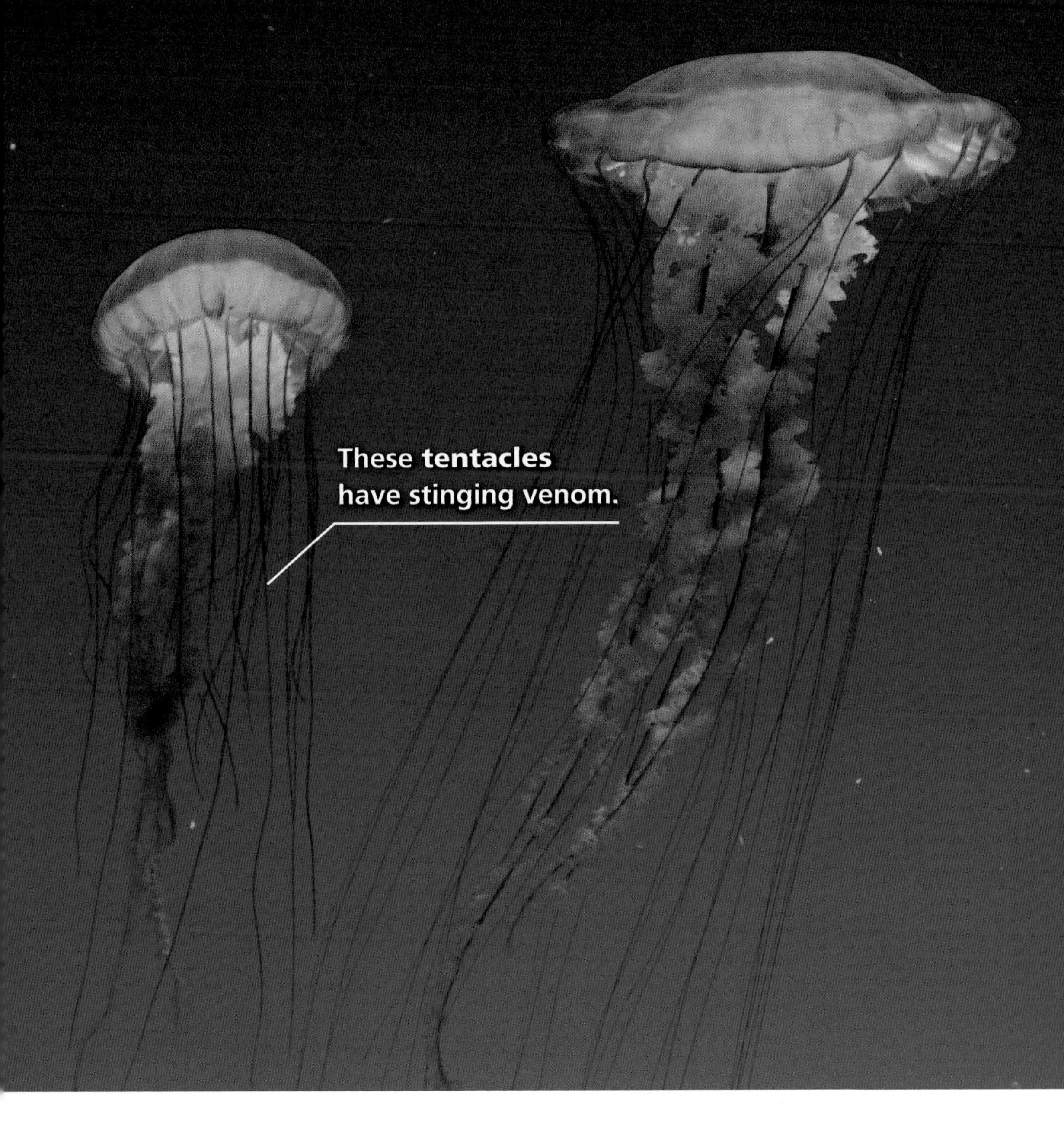

Most animal **plankton** are not very strong. They drift as the movement of the water pushes them. That is mainly how they move around the oceans.

A second group of plants and animals that live in the ocean, called the *benthos*, live on or in the ocean floor.

This group contains such unique life forms as coral, sponges, anemones (uh-NEM-uh-nees), **sea** stars, crabs, clams, and **sea** squirts. Corals and sponges are animals, but they do not have brains or **eyes**.

This is a **sea** star. Some people call it a starfish, but it is not a fish. Most sea stars have one tiny **eye** on the tip of each arm.

The giant clam seen in this picture has a body made up of two shells connected by large, strong muscles. That's why it's so hard to open a clamshell! A giant clam can weigh more than a gorilla and can live longer than people can.

Sally lightfoot crabs

Pygmy seahorse in a sea fan

This is a sea fan. It looks like a plant, but it is an animal. There is a sea horse hiding in the sea fan. The sea horse looks a lot like the sea fan.

Common octopus

The third major group of animals in the ocean is called *nekton*. These creatures swim freely through the water and include some of the most familiar of all sea life.

A few of the creatures that belong in this group are whales, sharks, manta rays, sea turtles, and well over 20,000 different species of fish.

Manta ray

There are many different kinds of sharks. The biggest is called the whale shark. It is the biggest fish in the ocean. It may be big, but it eats only tiny plankton.

Powder blue tang

Some kinds of fish swim together in large groups called ***schools***. These **schools** are usually made up of fish that are eaten as **prey** by larger fish. Maybe there really is safety in numbers!

Pygmy sweepers

Not all fish swim in **schools**. Some fish swim alone and hunt for other fish to eat. They often hunt at dusk, when it is hard for their **prey** to see them.

Most of the animals in the ocean are able to breathe underwater. However, some ocean animals need to come to the surface to breathe.

Whales, porpoises, and dolphins are mammals, just like us, and they would drown if they could not get to the surface for air.

Bottlenose dolphins

Dolphins like to be with other dolphins. They hunt and play together. They can see and hear very well, but they cannot smell.

Walrus

Sea lions

Sea lions, walruses, sea otters, and seals are mammals that spend much of their lives in the ocean. They might move slowly and clumsily on land, but they are swift and graceful in the water.

Sea lions

Sea otter

Most sea otters sleep on their backs in the water.
All sea otters eat on their backs in the water.
They like to eat fish, crabs, snails, and clams.

Sea otter

Sea turtle eggs

Sea turtle hatchling

Sea turtles spend most of their time underwater, only coming to the surface to breathe. Female turtles must leave the water and come ashore to **build** nests and lay their eggs. Then they quickly return to the water, leaving the eggs to hatch on their own.

Sea turtles **build** nests under the sand. After breaking out of its egg, a baby sea turtle must dig out of the sand. Then it must find its way to the ocean.

Brown pelican

Sea birds fly in the air above the ocean and feed on plants and animals from the ocean. They come in many different shapes and sizes. One of the most unusual birds is the **penguin**.

Gentoo penguins

Penguins do not fly in the **air** like most birds. They use their wings to help them swim. Some people say that they fly under the water!

CHAPTER 4 PEOPLE AND THE OCEAN

Positano, Italy

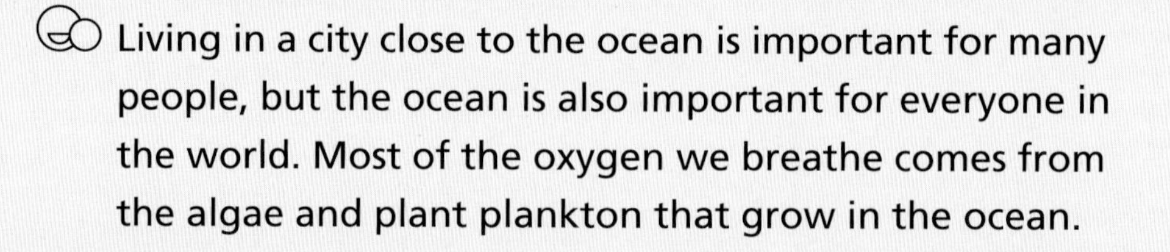

Living in a city close to the ocean is important for many people, but the ocean is also important for everyone in the world. Most of the oxygen we breathe comes from the algae and plant plankton that grow in the ocean.

San Francisco, California, USA

The ocean supports us in many ways. We get food from the ocean. Ships on the ocean can get us from place to place. We can even have fun at the beach!

Boy swimming with green sea turtle

People can have fun at an ocean beach. There they can swim, surf, snorkel, collect **seashells**, or just sit and listen to the **roar** of the waves. Once you have visited the ocean, you will want to return again and again!

At some time in the past, each **seashell** on the beach was part of a live sea animal. Some seashells are very small and some are very big. If you hold a big seashell to your ear, you might hear a sound like the soft **roar** of the ocean.

Snorkeling is a great way to see what's going on underwater. It's wonderful to discover such an amazing world lying just beneath the ocean's surface.

Many people fish to get food. Some people do it just because they like to. It's a nice way to spend time with friends and family.

Kelp seaweed forest

When most people think of the food that comes from the ocean, they think of fish. But there are many other kinds of food we harvest from the sea. In some countries, **seaweed** is used to make all kinds of delicious dishes!

Do you eat ice cream? Then you may have eaten **seaweed**. One kind of seaweed is often used to make ice cream thick.

Moving people and things across the vast ocean can be a real challenge. So people have built ships and boats of every size, from huge **freighters** to sleek sailboats to fancy cruise ships.

Container ship going under the Golden Gate Bridge

Freighters are big ships that move cargo from one place to another place. They can carry cars, food, toys, and even airplanes.

Scuba diver at shipwreck

Long ago big sailing ships would set out to **cruise** across the ocean. Occasionally one would end up sinking to the bottom of the sea. These old sunken ships can be found in places all across the oceans of the world. Some people try to find treasure in sunken ships. Do you think this one has any treasures on board?

These days, **cruise** ships carry people across the ocean. Some of these ships are like small towns. They have places to eat, shop, sleep, and just have fun.

CHAPTER 6 CARING FOR THE OCEAN

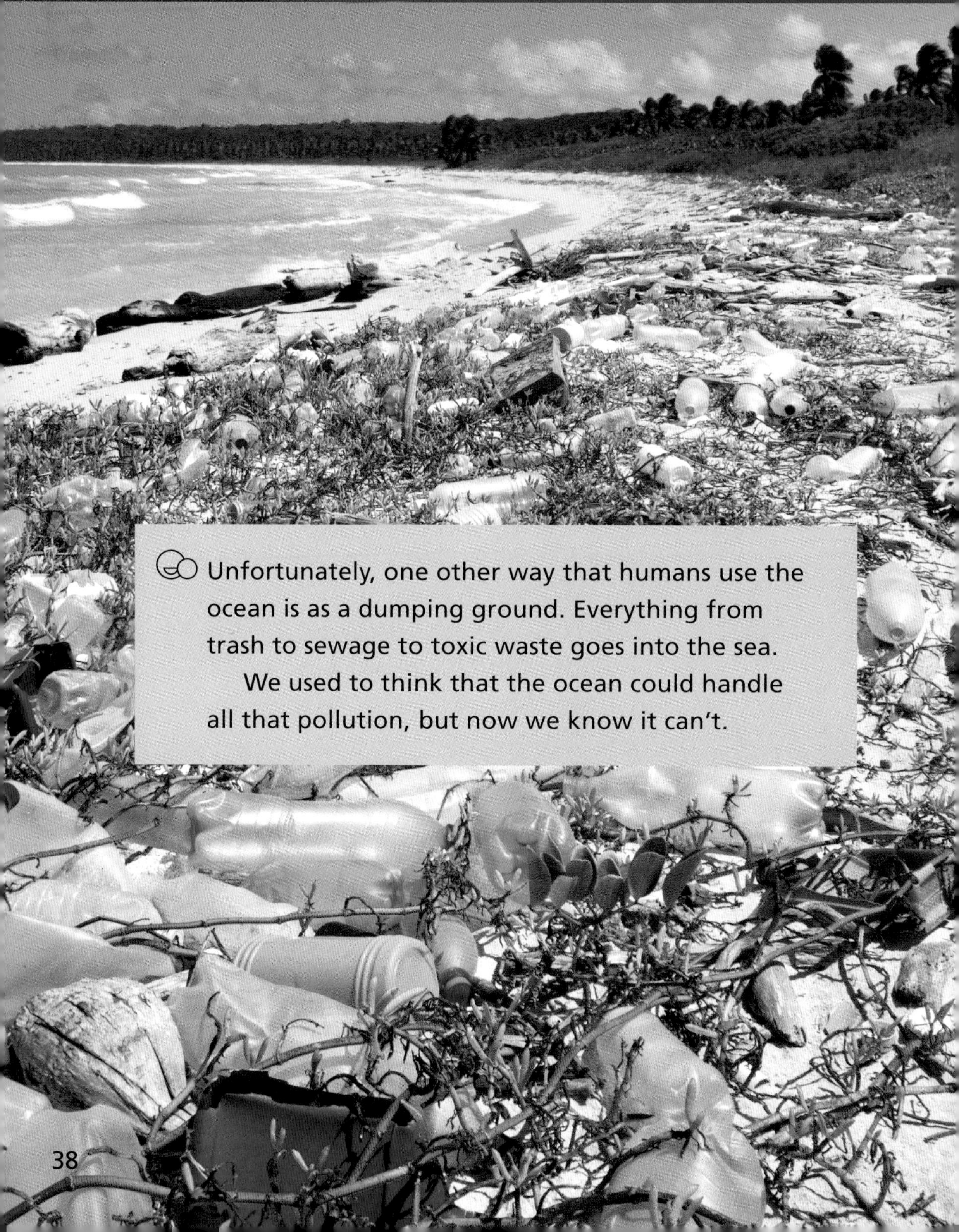

Unfortunately, one other way that humans use the ocean is as a dumping ground. Everything from trash to sewage to toxic waste goes into the sea.

We used to think that the ocean could handle all that pollution, but now we know it can't.

Big wave crashing

You can help. You can find out more about the ocean. You can share all you know with other people.

Picking up trash

The more you know about the ocean, the more you appreciate how important it is. It is one of our most precious resources. Life on Earth could not exist without it.

If we help take care of the ocean, the ocean will help take care of us.

Glossary

continent
one of the main landmasses on Earth

mammal
an animal that breathes air and has at least some hair or fur on its body

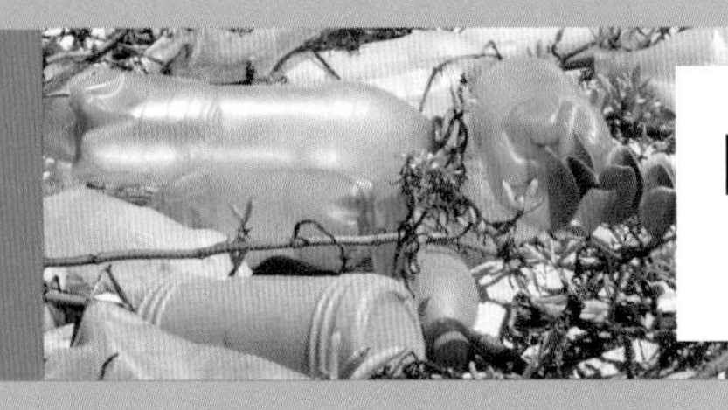

pollution
harmful substances in the air or water

predator
a fish or animal that hunts and eats other fish or animals

school
a group of fish swimming together

tide pool
a pool of water that remains after the tide goes out

Questions to Ask after Reading

Add to the benefits of reading this book by discussing answers to these questions. Also consider discussing a few of your own questions.

1. What fact in this book did you find the most interesting or surprising?
   Why was it interesting or surprising to you?

2. Was there a picture that you particularly liked?
   Why did you like that photograph?

3. What do you think would happen if small fish did not swim together in schools?
   Why do you think that might happen?

4. If you went to an ocean beach, what do you think you would like to do?

5. Why is the ocean important to all of us?
   Can you share three reasons?

If you liked ***The Ocean***, here are some other
We Both Read® books you are sure to enjoy!

You can see all the We Both Read books
that are available at WeBothRead.com.